tredition®

tredition was established in 2006 by Sandra Latusseck and Soenke Schulz. Based in Hamburg, Germany, tredition offers publishing solutions to authors and publishing houses, combined with worldwide distribution of printed and digital book content. tredition is uniquely positioned to enable authors and publishing houses to create books on their own terms and without conventional manufacturing risks.

For more information please visit: www.tredition.com

TREDITION CLASSICS

This book is part of the TREDITION CLASSICS series. The creators of this series are united by passion for literature and driven by the intention of making all public domain books available in printed format again - worldwide. Most TREDITION CLASSICS titles have been out of print and off the bookstore shelves for decades. At tredition we believe that a great book never goes out of style and that its value is eternal. Several mostly non-profit literature projects provide content to tredition. To support their good work, tredition donates a portion of the proceeds from each sold copy. As a reader of a TREDITION CLASSICS book, you support our mission to save many of the amazing works of world literature from oblivion. See all available books at www.tredition.com.

 Project Gutenberg

The content for this book has been graciously provided by Project Gutenberg. Project Gutenberg is a non-profit organization founded by Michael Hart in 1971 at the University of Illinois. The mission of Project Gutenberg is simple: To encourage the creation and distribution of eBooks. Project Gutenberg is the first and largest collection of public domain eBooks.

Many Ways for Cooking Eggs

Sarah Tyson Heston Rorer

Imprint

This book is part of TREDITION CLASSICS

Author: Sarah Tyson Heston Rorer
Cover design: Buchgut, Berlin – Germany

Publisher: tredition GmbH, Hamburg - Germany
ISBN: 978-3-8424-6301-1

www.tredition.com
www.tredition.de

Copyright:
The content of this book is sourced from the public domain.

The intention of the TREDITION CLASSICS series is to make world literature in the public domain available in printed format. Literary enthusiasts and organizations, such as Project Gutenberg, worldwide have scanned and digitally edited the original texts. tredition has subsequently formatted and redesigned the content into a modern reading layout. Therefore, we cannot guarantee the exact reproduction of the original format of a particular historic edition. Please also note that no modifications have been made to the spelling, therefore it may differ from the orthography used today.

CONTENTS

SAUCES

English Drawn Butter, Plain Hollandaise; Anchovy, Bechamel, Tarragon,
Horseradish, Cream or White, Brown Butter, Perigueux, Tomato, Paprika,
Curry, Italian

COOKING OF EGGS

To Preserve Eggs, Egging and Crumbing, Shirred Eggs, Mexicana, On a
Plate, de Lesseps, Meyerbeer, a la Reine, au Miroir, a la Paysanne, a la Trinidad, Rossini, Baked in Tomato Sauce, a la Martin, a la Valenciennes, Fillets, a la Suisse, with Nut-Brown Butter, Timbales, Coquelicot, Suzette, en Cocotte. Steamed in the Shell, Birds' Nests, Eggs en Panade, Egg Pudding, a la Bonne Femme, To Poach Eggs, Eggs
Mirabeau, Norwegian, Prescourt, Courtland, Louisiana, Richmond, Hungarian, Nova Scotia, Lakme, Malikoff, Virginia, Japanese, a la Windsor, Buckingham, Poached on Fried Tomatoes, a la Finnois, a la
Gretna, a l'Imperatrice, with Chestnuts, a la Regence, a la Livingstone, Mornay, Zanzibar, Monte Bello, a la Bourbon, Bernaise, a
la Rorer, Benedict, To Hard-boil, Creole, Curried, Beauregard, Lafayette, Jefferson, Washington, au Gratin, Deviled, a la Tripe, a l'Aurore, a la Dauphin, a la Bennett, Brouilli, Scalloped, Farci, Balls, Deviled Salad, Japanese Hard, en Marinade, a la Polonnaise, a la Hyde, a la Vinaigrette, a la Russe, Lyonnaise, Croquettes, Chops, Plain Scrambled, Scrambled with Chipped Beef, Scrambled with Lettuce,

Scrambled with Shrimps, Scrambled with Fresh Tomatoes, Scrambled with
Rice and Tomato, Scrambled with Asparagus Tips, Egg Flip

OMELETS

Omelet with Asparagus Tips, with Green Peas, Havana, with Tomato
Sauce, with Oysters, with Sweetbreads, with Tomatoes, with Ham, with
Cheese, with Fine Herbs, Spanish, Jardiniere, with Fresh Mushrooms,
O'Brien, with Potatoes

SWEET OMELETS

Omelet a la Washington, with Rum, Swiss Souffle, a la Duchesse, Souffle

SAUCES

The philosophy of a sauce, when understood, enables even an untrained cook to make a great variety of every day sauces from materials usually found in every household; to have them uniform, however, flavorings must be correctly blended, and measurements must be rigidly observed. Two level tablespoonfuls of butter or other fat, two level tablespoonfuls of flour, must be used to each half pint of liquid. If the yolks of eggs are added, omit one tablespoonful of flour or the sauce will be too thick. Tomato sauce should be flavored with onion, a little mace, and a suspicion of curry. Brown sauce may be simply seasoned with salt and pepper, flavored and colored with kitchen bouquet. Spanish sauce should also be flavored with mushrooms, or if you can afford it, a truffle, a little chopped ham, a tablespoonful of chives, shallot and garlic. Water sauce, drawn butter and simple sauce Hollandaise, when they are served with fish, must be flavored with a dash of tarragon vinegar, salt and pepper.

ENGLISH DRAWN BUTTER

3 tablespoonfuls of butter
1/2 pint of boiling water
2 tablespoonfuls of flour
1/2 teaspoonful of salt
1 dash of pepper

Rub two tablespoonfuls of butter and the flour together, add the boiling water, stir until boiling, add the salt and pepper; take from the fire, add the remaining tablespoonful of butter and it is ready for use. It must not be boiled after the last butter is added.

PLAIN SAUCE HOLLANDAISE

Make English drawn butter and add to it, when done, the yolks of two eggs beaten with two tablespoonfuls of water; cook until thick and jelly-like, take from the fire and add one tablespoonful of tarragon vinegar or the juice of half a lemon.

ANCHOVY SAUCE

Rub two teaspoonfuls of anchovy essence with the butter and flour and then finish the same as English drawn butter.

SAUCE BECHAMEL

2 tablespoonfuls of butter
1 yolk of an egg
1/2 cup of milk
1 saltspoonful of pepper
1 tablespoonful of flour
1/2 cup of stock
1/2 teaspoonful of salt

Rub the butter and flour together, add the stock and the milk and stir until boiling; add the salt and pepper, take from the fire and add the beaten yolk of the egg, heat for a moment over hot water, and it is ready for use.

TARRAGON SAUCE

Add two tablespoonfuls of tarragon vinegar to an English drawn butter.

HORSERADISH SAUCE

Make an English drawn butter, and, just at serving time, add a half cupful of freshly grated horseradish. If you are obliged to use that preserved in vinegar, press it perfectly dry before using it.

CREAM OR WHITE SAUCE

2 tablespoonfuls of butter
1/2 pint of milk
2 tablespoonfuls of flour
1/2 teaspoonful of salt
1 saltspoonful of pepper

Rub the butter and flour together, add the milk cold and stir until boiling; add the pepper and salt and it is ready for use.

BROWN BUTTER SAUCE

6 tablespoonfuls of butter 1 teaspoonful of mushroom catsup 1 tablespoonful of vinegar 4 tablespoonfuls of stock

Melt the butter, brown it and then skim; pour it carefully into a clean saucepan, add the vinegar, catsup and stock, boil a minute, and it is ready for use.

SAUCE PERIGUEUX

4 tablespoonfuls of butter
1/2 pint of stock
1 glass of white wine
1/2 teaspoonful of salt
2 tablespoonfuls of flour
1 bay leaf
2 chopped truffles
1 saltspoonful of pepper

1 teaspoonful of kitchen bouquet

Chop the truffles and put them with the bay leaf and wine in a saucepan on the back of the stove. Rub half the butter and flour together, add the stock, stir until boiling and add one teaspoonful of kitchen bouquet, the salt and pepper, and then the truffles; cook ten minutes, add the remaining quantity of butter and use at once.

TOMATO SAUCE

Rub together two level tablespoonfuls of flour and two of butter. Add a half pint of strained tomatoes. Stir until boiling. Add a teaspoonful of onion juice, a half teaspoonful of salt and a saltspoonful of pepper. Strain and use.

PAPRIKA SAUCE

Rub together two level tablespoonfuls of flour and two of butter, with a tablespoonful of paprika. Add a half pint of chicken stock. Stir until boiling. Add a half teaspoonful of salt, and strain. This sauce may be used over chicken as well as eggs.

CURRY SAUCE

Chop fine one onion. Cook it with two level tablespoonfuls of butter until soft. Do not brown. Add two level tablespoonfuls of flour, one teaspoonful of curry powder and a half teaspoonful of salt. Mix and add a half pint of boiling water. Stir until boiling, and strain.

ITALIAN SAUCE

Chop sufficient carrot to make a tablespoonful; chop one onion. Place them in a saucepan with three level tablespoonfuls of butter, a bay leaf and a blade of mace. Shake the pan over the fire until the vegetables are slightly browned. Drain off the butter and add to it two level tablespoonfuls of flour, a half cupful of good stock, a half cupful of strained tomatoes, and bring to a boil. Add a half teaspoonful of salt and a dash of cayenne. Strain. Stir until boiling, strain again and add four tablespoonfuls of sherry.

COOKING OF EGGS

Any single food containing all the elements necessary to supply the requirements of the body is called a complete or typical food. Milk and eggs are frequently so called, because they sustain the young animals of their kind during a period of rapid growth. Nevertheless, neither of these foods forms a perfect diet for the human adult. Both are highly nutritious, but incomplete.

Served with bread or rice, they form an admirable meal and one that is nutritious and easily digested. The white of eggs, almost pure albumin, is nutritious, and, when cooked in water at 170 degrees Fahrenheit, requires less time for perfect digestion than a raw egg. The white of a hard-boiled egg is tough and quite insoluble. The yolk, however, if the boiling has been done carefully for twenty minutes, is mealy and easily digested. Fried eggs, no matter what fat is used, are hard, tough and insoluble. The yolk of an egg cooks at a lower temperature than the white, and for this reason an egg should not be boiled unless the yolk alone is to be used.

Ten eggs are supposed to weigh a pound, and, unless they are unusually large or small, this is quite correct.

Eggs contain from 72 to 84 per cent. of water, about 12 to 14 per cent. of albuminoids. The yolk is quite rich in fat; the white deficient. They also contain mineral matter and extractives.

To ascertain the freshness of an egg without breaking it, hold your hand around the egg toward a bright light or the sun and look through it. If the yolk appears quite round and the white clear, it is fresh. Or, if you put it in a bucket of water and it falls on its side, it is fresh. If it sort of topples in the water, standing on its end, it is fairly fresh, but, if it floats, beware of it. The shell of a fresh egg looks dull and porous. As it begins to age, the shell takes on a shiny appearance. If an egg is kept any length of time, a portion of its water evaporates, which leaves a space in the shell, and the egg will "rattle." An egg that rattles may be perfectly good, and still not absolutely fresh.

TO PRESERVE EGGS

To preserve eggs it is only necessary to close the pores of the shells. This may be done by dipping them in melted paraffine, or packing them in salt, small ends down; or pack them in a keg and cover them with brine; or pack them in a keg, small ends down and cover them with lime water; this not only protects them from the air, but acts as a germicide.

Eggs should not be packed for winter use later than the middle of May or earlier than the first of April. Where large quantities of the yolks are used, the whites may be evaporated and kept in glass bottles or jars. Spread them out on a stoneware or granite plate and allow them to evaporate at the mouth of a cool oven. When the mixture is perfectly dry, put it away. This powder is capable of taking up the same amount of water that has been evaporated from it, and may then be used the same as fresh whites.

EGGS AND CRUMBING

To do this successfully one must prepare a mixture, and not use the egg alone. If an egg mixture or a croquette is dipped in beaten egg and rolled in cracker crumbs and dropped into fat, it always has a greasy covering. This is the wrong way. To do it successfully and have the articles handsome, beat the egg until well mixed, add a teaspoonful of olive oil, a tablespoonful of water and a dash of pepper. Dip the articles into this mixture, and then drop them on quite a thick bed of either sifted dry bread crumbs or soft white bread crumbs.

I prefer sifted dry bread crumbs for croquettes, and soft white crumbs for lobster cutlets and deviled crabs.

SHIRRED EGGS

Cover the bottoms of individual dishes with a little butter and a few fresh bread crumbs; drop into each dish two fresh eggs; stand this dish in a pan of hot water and cook in the oven until the whites are "set." Put a tiny bit of butter in the middle of each, and a dusting of salt and pepper.

EGGS MEXICANA

Put two tablespoonfuls of butter in a saucepan. Add four tablespoonfuls of finely chopped onion and shake until the onion is soft, but not brown. Then add four Spanish peppers cut in strips, a dash of red pepper and a half pint of tomatoes; the tomatoes should be in rather solid pieces. Add a seasoning of pepper and salt. Let this cook slowly while you shir the desired quantity of eggs. When the eggs are ready to serve, put two tablespoonfuls of this sauce at each side of the dish, and send at once to the table.

EGGS ON A PLATE

Rub the bottom of a baking dish with butter. Dust it lightly with salt and pepper. Break in as many fresh eggs as required. Stand the dish in a basin of water and cook in the oven five minutes, or until the whites are "set." While these are cooking, put two tablespoonfuls of butter in a pan and shake over the fire until it browns. When the eggs are done, baste them with the browned butter, and send to the table.

EGGS DE LESSEPS

Shir the eggs as directed. Have ready, carefully boiled, two sets of calves' brains; cut them into slices; put two or three slices between the eggs, and then pour over browned butter sauce.

EGGS MEYERBEER

To each half dozen eggs allow three lambs' kidneys. Broil the kidneys. Shir the eggs as directed in the first recipe. When done, put half a kidney on each side of the plate and pour over sauce Perigueux.

EGGS A LA REINE

 6 eggs
 1/2 pint of chopped cold cooked chicken
 1/2 can of mushrooms
 2 tablespoonfuls of butter
 2 tablespoonfuls of flour
 1/2 pint of milk
 1/2 teaspoonful of salt
 1 saltspoonful of pepper

Use ordinary shirring dishes for the eggs; butter them, break into each one egg, stand these in a pan of boiling water and in the oven until they are "set." Rub the butter and flour together, add the milk, stir until boiling, add the salt, pepper, chopped chicken and mushrooms, and put one tablespoonful of this on top of each egg and send at once to the table. This is also nice if you put a tablespoonful of the mixture in the bottom of the dish, break the egg into it, and then at serving time put another tablespoonful over the top.

EGGS AU MIROIR

Cover the bottom of a graniteware or silver platter with fresh bread crumbs, break in as many eggs as are needed for the number of persons to be served. Put bits of butter here and there, stand the platter over a baking pan of hot water in the oven until the eggs are "set," dust them with salt and pepper and send them to the table.

EGGS A LA PAYSANNE

6 eggs
1/2 cupful of cream
2 tablespoonfuls of grated onion
1 clove of garlic
1/2 teaspoonful of salt
1 saltspoonful of pepper

Add the onion and the garlic, mashed, to the cream; pour it in the bottom of a baking dish, break on top the eggs, dust with salt and pepper, stand the baking dish in a pan of water and cook in the oven until the eggs are "set." Serve in the dish in which they are cooked.

EGGS A LA TRINIDAD

6 eggs
2 lamb's kidneys
1 cupful of fresh bread crumbs
2 level tablespoonfuls of butter
2 level tablespoonfuls of flour
1/2 pint of stock
1 teaspoonful of kitchen bouquet
1/2 teaspoonful of salt
1 saltspoonful of pepper

Split the kidneys, cut out the tubes; scald them, drain, and cut them into thin slices. Put the butter into a saucepan, add the kidneys, toss until the kidneys are cooked, then add the flour, stock, kitchen bouquet, salt and pepper; stir until boiling. Grease a shallow granite or silver platter, break into it the eggs, sprinkle over the bread crumbs and stand them in the oven until the eggs are "set," then pour over the sauce, arrange the kidneys around the edge of the dish and send at once to the table.

EGGS ROSSINI

　6 eggs
　4 chicken livers
　12 nice mushrooms
　1/2 cupful of stock
　1/2 teaspoonful of salt
　1 dash of pepper

Put the stock in a saucepan and boil rapidly until reduced one-half, add a drop or two of browning. Throw the chicken livers into boiling water and let them simmer gently for ten minutes; drain. Slice the mushrooms and put them, with the livers, into the stock; let them stand until you have cooked the eggs. Put a tablespoonful of butter in the bottom of a shallow platter; when melted break in the eggs, stand them in the oven until "set," garnish with the livers and mushrooms and pour over the sauce.

EGGS BAKED IN TOMATO SAUCE

Make a tomato sauce. Pour one-half in the bottom of a baking dish or granite platter, break in from four to six fresh eggs, cover with the other half of the sauce, dust the top with grated cheese, and bake in a moderate oven until "set," about fifteen or twenty minutes. Serve for supper in the place of meat.

EGGS A LA MARTIN

Make a half pint of cream sauce. Put half of it in the bottom of a baking dish or into the bottom of ramekin dishes or individual cups. Break fresh eggs on top of the cream sauce, dust with a little salt and pepper, pour over the remaining cream sauce, sprinkle the top with grated cheese, and bake in a moderate oven until the cheese is browned and eggs are "set." Serve in the dish or dishes in which they are cooked.

EGGS A LA VALENCIENNE

6 eggs
1 pint of dry boiled rice
1/2 pint of strained tomato
2 mushrooms
2 tablespoonfuls of grated Parmesan cheese
2 level tablespoonfuls of butter
2 level tablespoonfuls of flour
1/2 saltspoonful of grated nutmeg
1/2 teaspoonful of paprika
1 teaspoonful of salt
1/2 saltspoonful of pepper

Rub the butter and flour together, add the strained tomato, stir until boiling, add the mushrooms, sliced, salt, paprika, nutmeg and pepper. Take a granite or silver platter, put in two tablespoonfuls of butter extra, let the butter melt and heat; break into this the eggs, being very careful not to break the yolks. Let the eggs cook in the oven until "set." Then put around the edge of the dish as a garnish the boiled rice, pour over the eggs the tomato sauce, dust the top with the Parmesan cheese and send at once to the table.

FILLETS OF EGGS

6 eggs
4 tablespoonfuls of good stock
1/2 teaspoonful of salt
1 saltspoonful of pepper

Beat the eggs with the stock, add the salt and pepper. Turn them into a buttered square pan, stand this in another of boiling water, and cook in the oven until the eggs are thoroughly "set." Cut the preparation into thin fillets or slices, dip in either a thin batter made from one egg, a half cupful of milk and flour to thicken, or they may be dipped in beaten egg, rolled in bread crumbs and fried in deep hot fat. Arrange the fillets in a platter on a napkin, one overlapping the other; garnish with parsley and send to the table with a boat of tomato or white sauce.

EGGS A LA SUISSE

Cover the bottom of a baking dish with about two tablespoonfuls of butter cut into bits. On top of this, very thin slices of Swiss cheese. Break over some fresh eggs. Dust with salt and pepper. To each half dozen eggs, pour over a half cup of cream. Then cover the top with grated Swiss cheese and bake in the oven until the cheese is melted and the eggs "set." Send this to the table with a plate of dry toast.

EGGS WITH NUT-BROWN BUTTER

These eggs may be shirred or poached and served on toast. Put two tablespoonfuls of butter in a saute or frying pan. As soon as it begins to heat, break into it the eggs and cook slightly until the yolks are "set;" dish them at once on toast or thin slices of broiled ham. Put two more tablespoonfuls of butter in the pan, let it brown, and add two tablespoonfuls of vinegar; boil it up once and pour over the eggs.

EGG TIMBALES

Butter small timbale molds or custard cups, dust the bottoms and sides with chopped tongue and finely chopped mushrooms. Break into each mold one fresh egg. Stand the mold in a baking pan half filled with boiling water, and cook in the oven, until the eggs are "set." Have ready nicely toasted rounds of bread, one for each cup, and a well-made tomato or cream sauce. Loosen the eggs from the cups with a knife, turn each out onto a round of toast, arrange neatly on a heated platter, fill the bottom of the platter with cream or tomato sauce, garnish the dish with nicely seasoned green peas and serve at once.

EGGS COQUELICOT

Grease small custard or timbale cups and put inside of each a cooked Spanish pepper. Drop in the pepper one egg. Dust it lightly with salt, stand the cups in a pan of boiling water and cook in the oven until the eggs are "set." Toast one round of bread for each cup and make a half pint of cream sauce. When the eggs are "set," fill the bottom of the serving platter with cream sauce, loosen the peppers from the cups and turn them out on the rounds of toast. Stand them in the cream sauce, dust on top of each a little chopped parsley and send to the table.

EGGS SUZETTE

Bake as many potatoes as you have persons to serve. When done, cut off the sides, scoop out a portion of the potato, leaving a wall about a half inch thick. Mash the scooped-out portion, add to it a little hot milk, salt and pepper, and put it into a pastry bag. Put a little salt, pepper and butter into each potato and break in a fresh egg. Press the potato from the pastry bag through a star tube around the edge of the potato, forming a border. Stand these in a baking pan and bake until the eggs are "set." Put a tablespoonful of cream sauce in the center of each, and send to the table.

EGGS EN COCOTTE

Chop fine one good-sized onion. Cook it, over hot water, in two level tablespoonfuls of butter. When the onion is soft add a quarter of a can of mushrooms, chopped fine, two level tablespoonfuls of flour and one cupful of stock. Stir until boiling. Add a tablespoonful of chopped parsley, a half teaspoonful of salt and a saltspoonful of pepper. Put a tablespoonful of this sauce in the bottom of individual cups. Break into each cup one egg. Pour over the remaining mixture. Stand the cups in a pan of hot water and bake in a moderate oven about five minutes.

EGGS STEAMED IN THE SHELL

Eggs put into hot water and kept away from the fire are much better than eggs actually boiled for only a short time. The greater the number of eggs to be cooked, the greater the amount of water that must be used. To cook four eggs, put them into a kettle, pour over them two quarts of water, cover the kettle and allow them to stand for ten minutes. Lift them from the water, put them into a large bowl, cover with boiling water, and send at once to the table. The whites will be coagulated, but should be soft and creamy, while the yolks will be perfectly cooked. If you should add six eggs to this volume of water, lengthen the time of standing. A single egg, dropped into a quart of water, must stand five minutes.

BIRDS' NESTS

Separate the eggs, allowing one to each person. Beat the whites to a stiff froth. Heap them into individual dishes, make a nest, or hole, in the center. Drop into this a whole yolk. Stand the dish in a pan of water, cover, and cook in the oven about two or three minutes. Dust lightly with salt and pepper, put a tiny bit of butter in the center of each, and send at once to the table. This is one of the most sightly of all egg dishes.

EGGS EN PANADE

2 eggs
6 slices of bread
1/2 cupful of milk or cream
4 tablespoonfuls of olive oil
1 tablespoonful of chopped parsley
1/2 teaspoonful of salt
1 saltspoonful of pepper

Trim the crusts from the bread. Beat the eggs until well mixed, but not light, then add the milk or cream, salt and pepper. Put the oil in a shallow frying pan, dip the slices of bread in the beaten egg and drop them into the hot oil; when brown on one side, turn and brown the other. Dish on a hot platter, dust with the chopped parsley and send at once to the table.

EGG PUDDING

6 eggs
6 slices of bread
1 tablespoonful of chopped parsley
2 tablespoonfuls of chopped chives
2 tablespoonfuls of butter
1 tablespoonful of flour
1/2 pint of milk
1/2 teaspoonful of salt
1 saltspoonful of white pepper

Break the eggs in a bowl, add all the seasoning. Rub the butter and flour together, add the milk, stir until boiling, and then add this to the eggs; beat together until thoroughly mixed. Crumb the bread, removing the crusts; stir this in at last. Turn into a buttered baking dish, cover with grated cheese, and bake in the oven until thoroughly "set" and a nice brown. It makes an exceedingly good, easily digested luncheon or supper dish for children.

EGGS A LA BONNE FEMME

1 Spanish or 2 Bermuda onions
2 level tablespoonfuls of butter
2 level tablespoonfuls of flour
1/2 pint of milk
6 eggs
1 teaspoonful of salt
1 saltspoonful of pepper
1/2 saltspoonful of grated nutmeg

Separate the whites and yolks of the eggs. Put the butter into a saucepan, add the onions, cut into *very thin* slices; shake until the onions are soft, but not brown, then dust over the flour, mix, and add the milk, salt, pepper and nutmeg. Stir carefully until this reaches boiling point, then stand it on the back part of the stove where it will keep hot for at least ten minutes. Beat the yolks of the eggs until very creamy, then stir them into the sauce, take from the fire, and fold in the well-beaten whites of the eggs. Turn into a baking dish or casserole and bake in a hot oven fifteen minutes; serve at once.

TO POACH EGGS

Use a shallow frying pan partly filled with boiling water. The eggs must be perfectly fresh. The white of an egg is held in a membrane which seems to lose its tenacity after the egg is three days old. Such an egg, when dropped into boiling water, spreads out; that is, it does not retain its shape. When ready to poach eggs, take the required number to the stove. The water must be boiling hot, but not actually bubbling. Break an egg into a saucer, slide it quickly into the water, and then another and another. Pull the pan to the side of the stove, where the water cannot possibly boil. With a tablespoon, baste the water over the yolks of the eggs, if they happen to be exposed. They must be entirely covered with a thin veil of the white. Have ready the desired quantity of toast on a heated platter, lift each egg with a slice or skimmer, trim off the ragged edges and slide them at once on the toast. Dust with salt and pepper, baste with melted butter, and send to the table.

EGGS MIRABEAU

Cut a sufficient number of rounds of bread, toast them carefully and cover them with *pate de foie gras*, put on top of each a poached egg, pour over sauce Perigueux, and send to the table.

EGGS NORWEGIAN

Cover rounds of toasted bread first with butter and then with anchovy paste, put on top of each a poached egg, pour over anchovy sauce, and send at once to the table.

EGGS PRESCOURT

Toast slices of bread, put thin slices of chicken on each, on top of this a poached egg, cover with sauce Bernaise, and serve at once.

EGGS COURTLAND

Mince sufficient cold chicken to make a half cupful. Make a half pint of cream sauce, add the minced chicken, a half teaspoonful of salt and a dash of red pepper. Toast a sufficient quantity of bread, put it on a heated platter, pour over a small quantity of the minced chicken and cream sauce, put on each a poached egg, cover with the remaining sauce, dust with parsley and serve with a garnish of green peas.

EGGS LOUISIANA

Make a half pint of tomato sauce, toast a sufficient quantity of bread, butter the bread and put on each slice a poached egg; cover with the tomato sauce.

EGGS RICHMOND

Chop sufficient cold chicken to make a half cupful, add an equal quantity of finely-chopped mushrooms, add this to a half pint of cream sauce. Add one unbeaten egg to a pint of cold boiled rice, season it with salt and pepper, make into round, flat cakes, and fry in hot fat. Arrange these on a heated platter, pour over the cream sauce mixture, and put on top of each a poached egg.

HUNGARIAN EGGS

Boil a cup of rice until tender and dry. Make a half pint of paprika sauce. Turn the rice into the center of a platter, smooth it down, cover the top with poached eggs, pour over the paprika sauce and send at once to the table.

EGGS NOVA SCOTIA

Put a poached egg on top of a flat codfish cake, pour over cream or tomato sauce, and send to the table.

EGGS LAKME

Cut cold chicken or turkey into very thin slices, and stand over hot water, in a dish, until heated; toast a sufficient quantity of bread, butter the slices, put on each a slice of chicken or turkey, dust lightly with salt and pepper. On top of these place a poached egg, cover with tarragon sauce, and send to the table.

EGGS MALIKOFF

Toast rounds of bread, cover them with caviar which has been seasoned with a little onion and pepper. Put on top of each a poached egg, cover with horseradish sauce, and send to the table.

EGGS VIRGINIA

Grate six ears of corn. Add half cupful of milk, a half cupful of flour and two eggs, beaten separately, and a half teaspoonful of salt and a dash of pepper. Drop the mixture in large tablespoonfuls in hot fat. When brown on one side, turn and brown on the other. Drain and arrange neatly on a large platter. Put a poached egg on the top of each cake, cover with cream sauce and send to the table. This dish, with green peas, makes quite a complete meal.

JAPANESE EGGS

Carefully boil one cup of rice, drain dry. Make a half pint of cream sauce, add to it a teaspoonful of grated onion and a teaspoonful of chopped celery. Poach the desired number of eggs. Put the rice in the center of a platter, cover it with the eggs, pour over the sauce. Dust the dish with parsley, and send at once to the table. The edge of this dish may be garnished with broiled sardines or carefully broiled smoked salmon.

EGGS A LA WINDSOR

  6 eggs
  6 rounds of toast
  2 level tablespoonfuls of butter
  2 level tablespoonfuls of flour
1/2 pint of chicken stock
  1 tablespoonful of chopped parsley
  1 tablespoonful of chopped olive
  1 tablespoonful of chopped Spanish pepper
1/2 teaspoonful of salt
  1 saltspoonful of black pepper

Rub the butter and flour together and add the stock; stir until boiling, and add the salt and pepper. Toast the bread. Poach the eggs, put them on the toast, pour over carefully the sauce, heap the chopped vegetables, mixed, in the center of each egg and send to the table.

EGGS BUCKINGHAM

Allow one egg to each person that is to be served. Cut either a dry or a Virginia ham into very thin slices; allow one thin square to each person. Toast squares of bread, remove the crust. Broil the ham quickly; put each square of ham on a square of toast, put on top a poached egg, dust lightly with pepper and send to the table.

POACHED EGGS ON FRIED TOMATOES

Cut solid tomatoes into slices a quarter of an inch thick, dust them with salt and pepper, dip them in egg beaten with a tablespoonful of water, roll them thickly with bread crumbs, dip them again in the egg, dust again with bread crumbs, and fry in deep hot fat. Drain on brown paper, dish on a heated platter, put a poached egg in the center of each slice, dust with salt and pepper, put a tablespoonful of tomato sauce over each egg and send at once to the table. Cream sauce may be used in the place of tomato sauce.

EGGS A LA FINNOIS

6 eggs
2 level tablespoonfuls of butter
2 level tablespoonfuls of flour
1/2 pint of strained tomato
1 tablespoonful of chopped chives
2 green peppers

Rub the butter and flour together, add the tomatoes, and the peppers, chopped very fine. Stir until this reaches boiling point, and stand it over hot water. Poach the eggs in deep water. Toast six rounds of bread; arrange the toast on a platter, put one egg on each slice, pour around the tomato sauce, dust thickly with the chives and send to the table.

EGGS A LA GRETNA

6 eggs
2 heads of celery
2 level tablespoonfuls of butter
2 level tablespoonfuls of flour
1/2 pint of milk
1 teaspoonful of salt
1 saltspoonful of pepper

Cut the celery into inch lengths, wash thoroughly, cover with boiling water and simmer gently thirty minutes until the celery is tender; drain, saving the water in which the celery was cooked for another purpose. Rub the butter and flour together, add the milk, salt and pepper; when boiling add the celery; stand this over hot water while you poach the eggs and toast six squares of bread. Butter the toast, put on each slice one egg; put these around the edge of a large platter, turn the celery into the middle of the dish and send at once to the table. To increase the beauty of this dish, and to give it a greater food value, you may garnish between the toast and celery with carefully boiled rice; this then makes an exceedingly nice supper dish.

EGGS A L'IMPERATRICE

Toast six slices of bread; butter them, put on top a thin slice of *pate de foie gras*, and on top of this a hot poached egg. Baste with a little melted butter, dust with salt and pepper and send at once to the table. This is one of the most elegant of all the egg dishes.

EGGS WITH CHESTNUTS

This is an exceedingly nice dish to serve in the Fall when chestnuts are fresh. Shell a quart of chestnuts, blanch them, then boil them until tender; drain and press through a colander. Add a half cupful of hot milk, a tablespoonful of butter, a teaspoonful of salt and a saltspoonful of pepper. Beat until light and stand over a kettle of hot water while you poach six or eight eggs. Dish the chestnut puree in a small platter, cover the poached eggs over the top, dust them with salt, pepper and chopped parsley.

EGGS A LA REGENCE

6 eggs
1/2 cupful of chopped cold cooked ham
1 grated onion
1/2 can of chopped mushrooms
2 tablespoonfuls of butter
2 tablespoonfuls of flour
1/2 pint of chicken stock
1/2 teaspoonful of salt
1 saltspoonful of pepper

Stand the ham over hot water until thoroughly heated. Rub the butter and flour together, add the stock, stir until boiling, add the mushrooms, sliced, the salt, pepper and the onion; stand this over hot water while you poach the eggs. Dish the eggs, cover them with the sauce, strained, and cover with the chopped ham. Garnish the dish with mashed potatoes or boiled rice, and send at once to the table.

EGGS A LA LIVINGSTONE

6 squares of toast
1 tureen of pate-de-foie-gras
6 eggs
1/2 cupful of good stock
2 tablespoonfuls of sherry
1 teaspoonful of kitchen bouquet
1/2 teaspoonful of salt
1 dash of pepper

Toast the bread, butter it and put on top of each slice of toast a slice of *pate de foie gras*; put this on a heated dish, stand it at the mouth of the oven door while you poach the eggs. Put into a saucepan all the other ingredients, bring to a boil, put one poached egg on each slice of *pate de foie gras*; baste with the sauce and send at once to the table.

EGGS MORNAY

6 eggs
2 tablespoonfuls of butter
2 tablespoonfuls of flour
1/2 pint of milk
1/2 teaspoonful of salt
1/2 teaspoonful of paprika
4 tablespoonfuls of grated Parmesan cheese

Rub the butter and flour together, add the milk, stir until boiling, add the salt and paprika, and if you have it, a teaspoonful of soy; pour half of this sauce in a shallow granite platter or baking dish. Poach the eggs, drain them carefully, and put them over the top of the sauce, cover with the remaining sauce, dust with Parmesan cheese and run in the oven a moment to brown.

EGGS ZANZIBAR

1 small egg plant
1 thin slice of ham
6 eggs
2 tablespoonfuls of sherry
2 tablespoonfuls of tomato catsup
2 level tablespoonfuls of butter
1 dash of pepper

Cut the egg plant into slices, season it with salt and pepper, dip in egg and bread crumbs and fry carefully in deep hot fat; put this on brown paper in the oven to dry. Broil the ham, cut it into squares sufficiently small to go neatly on top of each slice of egg plant. Poach the eggs, and heat the other ingredients for the sauce. Dish the egg plant on a platter, put on the ham, and on each piece of ham an egg; baste with sauce and send to the table.

EGGS MONTE BELLO

6 eggs
2 level tablespoonfuls of butter
2 level tablespoonfuls of flour
1/2 pint of strained tomato
1 teaspoonful of onion juice
1/2 teaspoonful of salt
1 saltspoonful of pepper

Put about two quarts of water into a small deep saucepan; when boiling very hard drop in, one at a time, the eggs. In dropping them in, the white will fold over the yolk and make the eggs round. Push them to the back of the stove to stand for two minutes. Lift them with a skimmer, dip them in an egg beaten with a tablespoonful of water, dust them with bread crumbs and fry them in deep hot fat. You cannot use a frying basket. Just drop them in the fat, and as they are browned lift them out onto soft paper to drain. Rub the butter and flour together, add the tomato and seasoning; when boiling dish the eggs on a heated platter, pour around tomato sauce and send to the table.

EGGS A LA BOURBON

6 eggs
1/2 pint of stock
1 tablespoonful of butter
6 tablespoonfuls of grated Parmesan cheese
1/2 teaspoonful of salt
1 dash of pepper

Put the stock in a small saucepan, poach the eggs in it, two at a time; lift them carefully and lay them on a hot granite or silver dish. When all are poached, dust over the cheese and stand them in the hot oven for just a moment until the cheese is melted. In the meantime boil the stock until it is reduced one-half, add the butter, baste it over the eggs and send to the table. This dish may be garnished with triangular pieces of toast.

EGGS BERNAISE

6 whole eggs
4 yolks of eggs
4 tablespoonfuls of stock
4 tablespoonfuls of olive oil
1 tablespoonful of chopped parsley
1 tablespoonful of tarragon vinegar
1 tablespoonful of butter
1 tablespoonful of flour

1/2 cupful of strained tomato
1 teaspoonful of onion juice
1/2 teaspoonful of salt

Put the stock, yolks of eggs and olive oil into a saucepan, stir over hot water until you have a thick, smooth sauce like mayonnaise; take from the fire, and when slightly cool stir in the tarragon vinegar and parsley. Rub the butter and flour together, add the tomato, and when boiling add a palatable seasoning of salt and pepper. Toast six halves of English muffins or squares of bread. Heat a platter, butter the toast, put it on the hot platter, and poach the eggs. Put one poached egg on each slice of toast, fill the bottom of the dish with tomato sauce and put a tablespoonful of Bernaise sauce on top of each egg. These may be garnished with a little chopped truffle, or a little chopped parsley.

EGGS A LA RORER

Toast rounds of bread, one for each person. Butter them. Heat, in boiling water, the choke of a French artichoke, one for each slice of bread. Make sauce Hollandaise, and put one artichoke bottom on each slice of bread on a heated platter. Put in the center a poached egg and pour over the sauce Hollandaise. Garnish the dish with nicely cooked French or fresh green peas.

EGGS BENEDICT

Separate two eggs. Break the yolks, add a cupful of milk, a half teaspoonful of salt, one and a half cupfuls of flour and a tablespoonful of melted butter. Beat well, add two level teaspoonfuls of baking powder and fold in the well-beaten whites. Bake on a griddle in large muffin rings. Broil thin slices of ham. Make a sauce Hollandaise. Chop a truffle. Poach the required number of eggs. Dish the muffins, put a square of ham on each, then a poached egg and cover each egg nicely with sauce Hollandaise. Dust with truffle and serve at once.

TO HARD-BOIL EGGS

Put the eggs in warm water, bring the water quickly to the boiling point, then push the kettle to the back of the stove, where the water will remain at 200 degrees Fahrenheit, for twenty minutes. If these are to be used for made-over dishes, throw them at once into cold water, remove the shells, or the yolks will lose their color.

EGGS CREOLE

Put two tablespoonfuls of butter and four of chopped onions into a saucepan, cook until the onion is soft, but not brown. Then add four peeled fresh tomatoes that have been cut into pieces, and three finely chopped green peppers. Cook this fifteen minutes, and add a level teaspoonful of salt. Have the eggs hard-boiled, and cut into slices. Put them into a baking dish, pour over the sauce, re-heat in the oven, and serve with a dish of boiled rice.

CURRIED EGGS

Peel, and cut into slices, three large onions. Put them in a saucepan with two tablespoonfuls of butter. Stand over hot water and cook until the onions are soft. Add a teaspoonful of curry powder, a clove of garlic mashed, a saltspoonful of ground ginger, a half teaspoonful of salt and a tablespoonful of flour; mix thoroughly and add a half pint of water. Stir until boiling. Have ready six hard-boiled eggs, cut them into slices, arrange them over a dish of carefully boiled rice, on a hot platter, strain over the sauce, and send at once to the table. This dish is made more attractive by a garnish with sweet Spanish peppers, cut into strips.

EGGS BEAUREGARD

Hard-boil five eggs. Separate the whites from the yolks. Put the yolks through a sieve. Put the whites either through a vegetable press, or chop them very fine. Make a half pint of cream sauce, season it and add the whites. Have ready a sufficient amount of toast, carefully buttered. Put this on a heated platter, cover over the cream sauce and the whites, dust the tops with the yolks, then with salt and pepper. Garnish the edge of the dish with finely chopped parsley, and send at once to the table.

EGGS LAFAYETTE

Hard-boil six eggs, chop them, but not fine. Make a half pint of curry sauce. Put the chopped eggs over a bed of carefully boiled rice, cover with the curry sauce, garnish with strips of Spanish pepper and serve. This dish may be changed by using tomato sauce in place of the curry sauce.

EGGS JEFFERSON

Select the desired number of good-sized tomatoes, allowing one to each person. Cut off the blossom end, scoop out the seeds, stand the tomatoes in a baking pan in the oven until they are partly cooked. Put a half teaspoonful of butter and a dusting of salt and pepper into the bottom of each, and break in one egg. Put these back in the oven until the eggs are "set." Have ready a round of toasted bread for each tomato, stand the tomato in the center of the bread, fill the bottom of the dish with cream sauce, and send to the table.

EGGS WASHINGTON

Add a half pint of crab meat to a half pint of cream sauce. Season with salt and pepper. Have ready either bread pates or pates made from puff paste. Put a tablespoonful of the crab mixture in the bottom of each. Break in an egg. Stand in the oven until the egg is "set." Or you may poach the eggs and slide them into the pate. Pour over the remaining quantity of crabmeat sauce, and send at once to the table.

EGGS AU GRATIN

Make a pint of cream sauce. Hard-boil six eggs. Cut them into slices. Put them in the baking dish and cover with the cream sauce. Dust thickly with cheese, and brown quickly in the oven.

DEVILED EGGS

Hard-boil twelve eggs. Remove the shells. Cut the eggs into halves, crosswise. Take out the yolks without breaking the whites. Press the yolks through a sieve. Add four tablespoonfuls of finely chopped chicken, tongue or ham. Add a half teaspoonful of salt, a saltspoonful of pepper and two tablespoonfuls of melted butter. Rub the mixture. Form it into balls the size of the yolks and put them into the places in the whites from which the yolks were taken. Put two halves together, roll them in tissue paper that has been fringed at the ends, giving each a twist. If these balls are made the size of the yolk, and put back into the whites, they may be placed on a platter, heated, and served on toast, with cream sauce; then they are very much like the eggs Bernhardt.

EGGS A LA TRIPE

Hard-boil eight eggs. Remove the shells, cut eggs crosswise in rather thick slices. Cut three small onions into very thin slices. Separate them into rings, cover them with boiling water and boil rapidly ten minutes; drain, then cover them with fresh water and boil until they are tender; drain again, but save the water. Now mix the eggs and onions carefully, without breaking. Put two level tablespoonfuls of butter and two of flour into a saucepan. Mix. Add a grating of nutmeg, a saltspoonful of black pepper, the juice of a lemon, and a half-pint of the water in which the onions were boiled. Bring to the boiling point, add two tablespoonfuls of cream; then add the eggs and onions. When thoroughly hot, dish them in a conical form, garnish with triangular pieces of toast, and serve.

EGGS A L'AURORE

Hard-boil six eggs, cut them into halves lengthwise, take out the yolks, keeping them whole. Cut the whites into fine strips. Make a cream sauce. Add to it two tablespoonfuls of finely chopped sardines or finely chopped lobster or crab, a tablespoonful of tarragon vinegar. Add the whites of the eggs, and, when quite hot, add the yolks, without breaking them. Turn this at once into a heated dish, garnish the dish with triangular pieces of toast, and send to the table. Or, if you like, make the sauce, season it and put a layer into the bottom of the baking-dish, then a layer of Parmesan cheese, then a layer of the yolks, pressed through a sieve, and so on, alternating, having the last layer of the yolks of the eggs. Dust over a few bread crumbs, put here and there bits of butter, and brown quickly in the oven.

EGGS A LA DAUPHIN

Remove the shells from six hard-boiled eggs, cut them into halves, lengthwise, take out the yolks, press them through a sieve. Add four level tablespoonfuls of melted butter, and half a teaspoonful of salt, a grating of nutmeg and two tablespoonfuls of Parmesan cheese. Add half a cupful of cream to a half cupful of sifted bread crumbs. Mix this with the yolks, rub until smooth, then add one well-beaten egg, and the yolk of one egg. Cover the bottom of the baking dish with the mixture forming it in a pyramid and cover with the chopped whites. Have ready two extra hard-boiled eggs, take out the yolks, press them through a sieve, all over the top. Garnish the edges of the dish with triangular pieces of toasted bread, cover the whole with cream sauce, brown in the oven, and serve at once.

EGGS A LA BENNETT

6 hard-boiled eggs
2 tablespoonfuls of butter
1 teaspoonful of anchovy sauce
1 tablespoonful of finely chopped chives or onion

1/2 cupful of bread crumbs
1/2 teaspoonful of salt

Cut the eggs into halves lengthwise; remove the yolks, rub them with half the butter, salt, onion and anchovy paste. Fill these back into the whites. Cover the bottom of a baking dish with ordinary white sauce, stand in the eggs, put over the bread crumbs, baste them with the remaining butter, melted, and stand in the oven long enough to brown.

EGGS BROULLI

Beat four eggs. Add to them four tablespoonfuls of stock, four tablespoonfuls of cream, a saltspoonful of salt and half a saltspoonful of pepper. Turn them into a saucepan, stand in a pan of hot water, stir with an egg-beater until they are thick and jelly-like. Turn at once into a heated dish, garnish with toast and send to the table.

SCALLOPED EGGS

4 hard-boiled eggs
2 tablespoonfuls of butter
2 level tablespoonfuls of flour
1/2 pint of milk
1 cupful of finely chopped cold cooked chicken or fish
1 teaspoonful of salt
1 saltspoonful of pepper

Chop the eggs rather fine. Rub the butter and flour together, add the milk, stir until boiling, add the salt and pepper. Put a layer of eggs in the bottom of a casserole, or baking dish, then a layer of the fish or chicken, then a little white sauce, and so continue until the ingredients are used. Dust the top thickly with bread crumbs and bake in a moderate oven until nicely browned.

EGG FARCI

6 hard-boiled eggs 2 cupfuls of mashed potatoes 1 cupful of finely chopped cold cooked meat 1 tablespoonful of chopped parsley 1 tablespoonful of butter 1 tablespoonful of flour 1 gill (a half cupful) of milk 1 level teaspoonful of salt 1 teaspoonful of onion juice 1 saltspoonful of pepper

Hard-boil the eggs, chop them fine, mix them with the meat, add the salt, pepper and parsley. Rub the butter and flour together, add the milk, stir until boiling; add this gradually to the potatoes. When smooth add the hard-boiled eggs, meat and parsley. Fill into small custard cups or into shirring dishes, brush with milk and brown in the oven. These make a nice supper or luncheon dish.

EGG BALLS

These are used for soup and for garnishing of vegetable dishes. Hard-boil four eggs, throw them at once into cold water, remove the shells. Put the yolks through a sieve, then add a half teaspoonful of salt, a dash of white pepper and the yolk of one raw egg, or you may take a part of the white of one egg. Mix thoroughly and make into balls the size of a marble, using enough flour to prevent sticking to the hands. Drop these into a kettle of boiling stock, or into hot fat. Drain on brown paper.

DEVILED EGG SALAD

6 eggs
1 head of lettuce
1 pimiento
1 teaspoonful of onion juice
1/2 teaspoonful of paprika
1/2 cupful of chopped boiled tongue
1 saltspoonful of salt
1 saltspoonful of pepper

Hard-boil the eggs, throw them into cold water, remove the shells, cut them lengthwise. Take out the yolks without breaking the whites. Rub the yolks through a sieve into a bowl, then add the tongue and all the seasoning. If the mixture is dry add a tablespoonful or two of cream or olive oil. Roll the mixture into balls that will fit the spaces from which they were taken in the whites, making each ball round. Arrange the lettuce over a platter, stand the whites in the lettuce, and at serving time baste thoroughly with French dressing.

JAPANESE HARD EGGS

1 cupful of rice
1/2 pint of white sauce
6 eggs
1 tablespoonful of chopped parsley, if you have it, and a suspicion of onion juice

Put the eggs into a saucepan of cold water, bring to boiling point, and simmer gently twenty minutes. Wash the rice through several cold waters, sprinkle it into a kettle of boiling water and boil it for thirty minutes. Remove the shells, break the eggs while they are hot, cut them into halves crosswise. Make the cream sauce, and add the onion juice. When the rice is done, drain, sprinkle it in the center of a large platter, press the halves of the eggs down into it, and pour over the cream sauce. Garnish with the chopped parsley. This takes the place of both meat and starchy vegetables for either luncheon or supper.

EGGS EN MARINADE

1 dozen eggs
3 very red beets
1 quart of cider vinegar
24 whole cloves
1 teaspoonful of mustard seed
1 saltspoonful of celery seed

1 teaspoonful of salt
2 saltspoonfuls of pepper

Hard-boil the eggs; plunge them into cold water and remove the shells. Stick the cloves into the eggs. Pare the beets, cut them into blocks and boil them in about a pint of water. To this water add the vinegar, bring it to boiling point, add salt, pepper and the celery and mustard seed. Put the eggs into a glass jar and pour over the boiling vinegar; put on the tops and stand them aside for three weeks. A tablespoonful of grated horseradish or a half cupful of nasturtium seeds will improve the flavor and prevent mold.

EGGS A LA POLONAISE

6 eggs
2 level tablespoonfuls of butter
1 tablespoonful of chopped parsley
1 teaspoonful of salt
1 saltspoonful of pepper

Hard-boil four of the eggs; when done remove the shells, cut the eggs into halves lengthwise and take out the yolks, without breaking the whites. Press the yolks through a sieve into a bowl, and add the raw yolks of the remaining two eggs, with the parsley, salt and pepper. Beat the white of the raw eggs until light, not stiff, then work them into the yolk mixture. Cover the bottom of a shallow baking pan with part of this mixture, then fill the spaces in the whites with some of the remaining mixture. Put the whites of the eggs together, making them look like whole eggs. Arrange these in the center of the dish. If you have any of the yolk mixture left, put it around in a sort of a border. Pour over a little melted butter, dust thickly with soft bread crumbs and bake in a quick oven until slightly brown. Serve plain or with cream sauce.

EGGS A LA HYDE

6 eggs
1/2 can of mushrooms
1 tablespoonful of grated onion
2 tablespoonfuls of chopped parsley
1/2 cupful of sweet cream
2 level tablespoonfuls of butter
2 level tablespoonfuls of flour
1/2 pint of chicken stock or cocoanut milk
1 teaspoonful of salt
1 saltspoonful of pepper

Hard-boil the eggs, and when done remove the shells and cut the eggs into halves lengthwise, keeping the whites whole. Remove the yolks, press them through a sieve, add to them the cream, half the salt and a dash of cayenne. Mix thoroughly and fill into the whites and arrange them neatly on a granite or silver platter. Put the butter into a saucepan, add the onion and flour, then the stock or cocoanut milk, and the mushrooms; stir, until it boils, add the remaining salt and pepper; take from the fire and add the parsley. Pour this over the eggs on the platter, dust thickly with bread crumbs, run into a quick oven until brown.

EGGS A LA VINAIGRETTE

6 eggs
1 head of lettuce
8 tablespoonfuls of olive oil
1 tablespoonful of chopped parsley
4 tablespoonfuls of vinegar
1 tablespoonful of chopped gherkin
1 tablespoonful of chopped olives
1 tablespoonful of grated onion

Hard-boil the eggs, throw them into cold water; remove the shells and cut them into slices lengthwise. Wash and dry the lettuce, arrange it on a small meat platter, put over the top slices of hard-boiled eggs, letting one slice overlap the other. Fill the center of the dish with sliced, peeled tomatoes. Put a half teaspoonful of salt in a soup plate, add a saltspoonful of pepper and the oil; put in a piece of ice and stir until the salt is dissolved. Remove the ice, add all the other ingredients but the parsley, mix thoroughly, pour this over the eggs, dust with parsley and serve as a supper dish.

EGGS A LA RUSSE

6 eggs
1 small can of caviar (2 tablespoonfuls)
1/2 pint of stock
1 teaspoonful of onion juice
1 dash of pepper

Hard-boil the eggs, remove the shells, cut them into halves lengthwise; take out the yolks without breaking the whites, and press them through a sieve, then add the caviar, onion juice and pepper. Heap these back into the whites. Boil the stock until reduced one-half, baste the eggs carefully, run them into the oven until hot, pour over the remaining hot stock and send to the table.

EGGS LYONNAISE

6 eggs
1 onion
2 level tablespoonfuls of butter
2 level tablespoonfuls of flour
1/2 pint of milk
1/2 teaspoonful of salt
1 dash of pepper

Hard-boil the eggs, remove the shells, throw them in cold water. Cut the onion into thin slices; put it, with the butter, into a saucepan, shake until the onion is tender, then add the flour, milk and seasoning; stir until boiling. At serving time cut the eggs into slices crosswise, put them in a shallow baking dish, cover with cream sauce and run in the oven just a moment until they are very hot.

EGG CROQUETTES

6 eggs
1/2 pint of milk
2 level tablespoonfuls of butter
3 level tablespoonfuls of flour
1 teaspoonful of onion juice
1 tablespoonful of chopped parsley
1/2 saltspoonful of grated nutmeg
1 teaspoonful of salt
1 saltspoonful of pepper

Hard-boil the eggs and chop them fine. Rub the butter and flour together, add the milk, stir until you have a thick, smooth paste. Add all the seasoning to the egg, mix the eggs into the white sauce and turn out to cool. When cold form into cylinders, dip in egg beaten with a tablespoonful of water, roll in bread crumbs and fry in deep hot fat. Serve with cream sauce.

EGG CHOPS

6 hard-boiled eggs
1/2 pint of finely chopped cooked ham
1/2 pint of milk
2 level tablespoonfuls of butter
4 level tablespoonfuls of flour
1 tablespoonful of chopped parsley
1 teaspoonful of onion juice
1/2 teaspoonful of salt
1 dash of cayenne
1 dash of white pepper

Chop the eggs very fine, mix them with the ham; add the parsley, onion juice and pepper. Rub the butter and flour together and add the milk. Stir until you have a smooth, thick sauce, then add the salt; mix this with the other ingredients and turn it out to cool. When cold form into a chop about the size of an ordinary mutton chop. Dip first in egg beaten with a tablespoonful of water, then cover carefully with bread crumbs and fry in deep hot fat. Serve with either tomato or brown sauce.

PLAIN SCRAMBLED EGGS

Put two tablespoonfuls of butter in a shallow frying pan. Add a tablespoonful of water to each egg. Six eggs are quite enough for four people. Add a half teaspoonful of salt, and a saltspoonful of pepper. Give two or three beats—enough to break the eggs; turn them into the frying pan, on the hot butter. Constantly scrape from the bottom of the pan with a fork, while they are cooking. Serve with a garnish of broiled bacon and toast.

SCRAMBLED EGGS WITH CHIPPED BEEF

Pull apart a quarter of a pound of chipped beef, cover with boiling water, let it stand ten minutes, drain and dry. Put it into a saucepan with two level tablespoonfuls of butter, four eggs, beaten until they are well mixed, and a dash of pepper. Stir with a fork until the eggs are "set."

EGGS SCRAMBLED WITH LETTUCE

Remove the outside leaves from one head of lettuce; wash, dry, and with a very sharp knife cut them into shreds. Chop sufficient onion to make a tablespoonful. Put a tablespoonful of butter into a saucepan, add the onion, shake until the onion is soft, then add six eggs, beaten without separating until well mixed, but not light. Add a half teaspoonful of salt, a half saltspoonful of pepper and the shredded lettuce. Stir with a fork until the eggs are "set," turn at once onto a heated platter, garnish with triangular pieces of toast and send to the table.

SCRAMBLED EGGS WITH SHRIMPS

6 eggs
1 can of shrimps or its equivalent in fresh shrimps
1 green pepper
1/2 pint of strained tomato
1/2 teaspoonful of salt

Beat the eggs until well mixed, without separating. Put the butter in a saucepan, add the pepper, chopped; shake until the pepper is soft, add the tomato and all the seasoning, and the shrimps. Bring to boiling point, push to the back of the stove where it will simmer while you scramble the eggs. Put the scrambled eggs on toast in the center of a platter, pour over and around the shrimp mixture and send to the table.

EGGS SCRAMBLED WITH FRESH TOMATOES

3 tomatoes
4 eggs
1 teaspoonful of onion juice
1 level teaspoonful of salt
1 saltspoonful of pepper
2 tablespoonfuls of butter

Peel the tomatoes, cut them into halves and squeeze out the seeds. Cut the tomatoes into small bits, put them into a saucepan with the salt, pepper and butter; when these are hot add the eggs, beaten until well mixed, stir until the eggs are "set," turn into a heated dish, garnish with toast and send to the table.

EGGS SCRAMBLED WITH RICE AND TOMATO

This is an exceedingly nice dish for supper where one does not care for meat. Four or six eggs can be used to each half-pint of cold boiled rice, and either three fresh tomatoes, chopped, or two-thirds of a cupful of solid strained tomato. Put a tablespoonful of butter, a half teaspoonful of salt, a saltspoonful of pepper and the tomatoes into a saucepan; when hot add the rice, and when the rice is hot add the eggs, beaten without being very light. As soon as the eggs are "set" serve this in a vegetable dish covered with squares of toasted bread. This recipe is also nice with hard-boiled eggs; proceed as directed, and at last add the hard-boiled eggs, sliced.

SCRAMBLED EGGS WITH ASPARAGUS TIPS

 1 small can of asparagus tips
 6 eggs
 1 tablespoonful of butter
 1/2 teaspoonful of salt
 1 dash of pepper

Beat the eggs, add the salt, pepper and butter. Put them into a saucepan, add at once the asparagus tips and stir with a fork until the mixture is "set."

EGG FLIP

This dish is exceedingly nice for a child or an invalid. Separate one egg, beat the white to a stiff froth, add the yolk and beat again. Heap this in a pretty saucer, dust lightly with powdered sugar, put in the center a teaspoonful of brandy, and serve at once. Sherry or Madeira may be substituted for the brandy.

OMELETS

A plain French omelet is, perhaps, one of the most difficult of all things to make; that is, it is the most difficult to have well made in the ordinary private house. Failures come from beating the eggs until they are too light, or having the butter too hot, or cooking the omelet too long before serving.

In large families, where it is necessary to use a dozen eggs, two omelets will be better than one. A six-egg omelet is quite easily handled. Do not use milk; it toughens the eggs and gives an unpleasant flavor to the omelet. An "omelet pan," a shallow frying pan, should be kept especially for omelets. Each time it is used rub until dry, but do not wash. Dust it with salt and rub it with brown paper until perfectly clean.

To make an omelet: First, put a tablespoonful of butter in the middle of the pan. Let it heat slowly. Break the eggs in a bowl, add a tablespoonful of water to each egg and give twelve good, vigorous beats. To each six eggs allow a saltspoonful of pepper, and, if you like, a tablespoonful of finely chopped parsley. Take the eggs, a limber knife and the salt to the stove. Draw the pan over the hottest part of the fire, turn in the eggs, and dust over a half teaspoonful of salt. Shake the pan so that the omelet moves and folds itself over each time you draw the pan towards you. Lift the edge of the omelet, allowing the thin, uncooked portion of the egg to run underneath. Shake again, until the omelet is "set." Have ready heated a platter, fold over the omelet and turn it out. Garnish with parsley, and send to the table.

If one can make a plain French omelet, it may be converted into many, many kinds.

OMELET WITH ASPARAGUS TIPS

Make a plain omelet from six eggs, have ready a half pint of cream sauce, and either a can or a bundle of cooked asparagus. Cut off the tips, preserving the lower portions for another dish. When the omelet is turned onto the heated platter, put the asparagus tips at the ends, cover them with cream sauce, pour the rest of the cream sauce in the platter, not over the omelet.

OMELET WITH GREEN PEAS

Make a six-egg omelet. Have ready one pint of cooked peas, or a can of peas, seasoned with salt, pepper and butter. Just before folding the omelet put a tablespoonful of peas in the center, fold, and turn out on a heated platter. Pour the remaining quantity of peas around the omelet, and send at once to the table. If you like, you may pour over, also, a half pint of cream sauce.

HAVANA OMELET

Put two tablespoonfuls of butter and two chopped onions over hot water until the onion is soft and thoroughly cooked. Peel four tomatoes, cut them into halves and press out the seeds. Then cut each half into quarters, add four Spanish peppers cut in strips, a level teaspoonful of salt and a dash of red pepper. Cook until the tomato is soft. Make a six-egg omelet. Turn it onto a heated platter, put the tomato mixture at the ends, and send at once to the table.

OMELET WITH TOMATO SAUCE

Make a plain omelet with six eggs. Pour over a half pint of tomato sauce, and send to the table.

OMELET WITH OYSTERS

Drain, wash, and drain again twenty-five oysters. Throw them into a hot saucepan and shake until the gills curl. Rub together two level tablespoonfuls of flour and two of butter. Drain the oysters, put the liquor into a half-pint cup, add sufficient milk to fill the cup. Add this to the butter and flour. When boiling, add the oysters, a level teaspoonful of salt and a dash of red pepper. Make a six-egg omelet, turn it onto a heated dish, arrange the oysters around the omelet, pour over the cream sauce, and send to the table.

OMELET WITH SWEETBREADS

This is a very good way to make sweetbreads do double duty. Boil a pair of sweetbreads until they are tender. Remove the membrane, cut them into slices; make a cream sauce. Add the sweetbreads, and, if you like, a half can of chopped mushrooms. Make a six-egg omelet, arrange the slices of sweetbread around the omelet and pour over the cream sauce.

OMELET WITH TOMATOES

Beat six eggs. Add a half pint of rather thick stewed tomatoes, a level teaspoonful of salt and a saltspoonful of pepper. Beat the eggs and tomatoes together, and make precisely the same as a plain omelet. Do not, however, add water, as the tomatoes answer the purpose.

OMELET WITH HAM

Mix a half cup of chopped ham with the eggs after they have been beaten with the water, and finish the same as a plain omelet.

OMELET WITH CHEESE

Beat six eggs until they are thoroughly mixed. Add a half cupful of thick cream, four tablespoonfuls of grated cheese, a saltspoonful of black pepper and a half teaspoonful of salt. Mix and finish the same as plain omelet.

OMELET WITH FINE HERBS

Beat six eggs until thoroughly mixed. Add a half cupful of cream, a tablespoonful of finely chopped parsley, a saltspoonful of pepper and a half teaspoonful of salt. Finish the same as a plain omelet. Serve on a heated platter and put over a little thin Spanish sauce.

SPANISH OMELET

Beat six eggs. Add six tablespoonfuls of water. Add a saltspoonful of pepper, a tablespoonful of finely chopped parsley, a teaspoonful of onion juice. Put six thin slices of bacon in the omelet pan. Cook slowly until all the fat is tried out. Remove the bacon, add a tablespoonful of chopped onion. Cook until the onion is slightly brown, turn in the eggs and finish the same as a plain omelet. Turn onto a heated platter, garnish with red and green peppers, and, if you like, put two tablespoonfuls of stewed tomatoes at each end of the omelet.

OMELET JARDINIERE

Chop sufficient chives to make a tablespoonful. Add a tablespoonful of parsley, a tablespoonful of finely chopped onion, and, if you have it, a little of the green tops of celery. Mix this with six eggs, add six tablespoonfuls of water and beat. Make the same as a plain omelet.

OMELET WITH FRESH MUSHROOMS

This is one of the most delicious of all the luncheon dishes. Put two tablespoonfuls of butter, a pound of mushrooms, sliced, a half cup of milk and a teaspoonful of salt into a saucepan. Cover and cook slowly for twenty minutes. Make two six-egg omelets. Turn them, side by side, on a large heated platter, pour over the fresh mushrooms and serve at once.

OMELET O'BRIEN

Put two tablespoonfuls of butter in a saucepan with four tablespoonfuls of chopped onion. Cook until the onion is tender. Then add four chopped Spanish peppers, two tablespoonfuls of thick tomato, or one whole raw tomato cut into bits, four sliced cooked okra, a teaspoonful of salt, a dash of pepper. Let these cook twenty minutes. Make a six-egg plain omelet, using bacon fat instead of butter for the cooking. Remove the slices of bacon before they are too hard, as they must be used for a garnish. Turn the omelet onto a heated platter, pour around it the pepper mixture, garnish with the bacon, and send to the table. Canned mushrooms may be added, if desired.

OMELET WITH POTATOES

4 eggs
1 cupful of mashed potatoes
2 level tablespoonfuls of butter
1 tablespoonful of chopped parsley
1 level teaspoonful of salt
1 saltspoonful of pepper

Beat the eggs, without separating, until thoroughly mixed; add them gradually to the mashed potato, beating all the while; add the salt and pepper. Put the butter into a good-sized saute or omelet pan; when hot, turn the ingredients into the pan, and smooth it down with a pallet knife. Let this cook slowly until nicely browned; fold it over as you would a plain omelet, and turn onto a heated dish. The parsley may be sprinkled over the top, or added to the mixture.

SWEET OMELETS

OMELET A LA WASHINGTON

Put three eggs into a bowl, and three into another bowl. Add three tablespoonfuls of water to each, and beat. Have two omelet pans, in which you have melted butter. Grate an apple into one bowl, and into the other put a little salt and pepper. Stand two tablespoonfuls of jelly in a dish over hot water while you cook the omelets. Proceed as for plain omelet. The one to which you have added the apple, turn out on a plate. Before folding the other, put in the center the softened currant jelly, then fold it and turn it out by the side of the other omelet. Dust both with powdered sugar, and send at once to the table. Serve a portion of each.

OMELET WITH RUM

Make a plain omelet with six eggs, turn it on a heated platter. Dust it with powdered sugar, and score it across the top with a red-hot poker. Dip four lumps of sugar into Jamaica rum and put them on the platter. Put over the omelet four tablespoonfuls of rum; touch a lighted match to the rum, and carry the omelet to the table, burning. Baste it with the burning rum until the alcohol is entirely burned off.

SWISS SOUFFLE

Allow one egg to each person. Have everything in readiness. The maraschino cherries must be drained free from the liquor. Separate the eggs. Beat the whites until they are stiff. Add a level tablespoonful of powdered sugar to each white, and beat until dry and glossy. Add the yolks of three eggs. Mix quickly. Add the grated rind of one lemon and a tablespoonful of lemon juice. Heap this into individual dishes. Make a tiny little hole in the center and put in a maraschino cherry, leaving the hole large enough to hold a tablespoonful of the liquor when the omelet is ready to serve; dust it with powdered sugar, bake in a quick oven about three minutes, take it from the oven, pour in the maraschino juice and send *at once* to the table. These will fall if baked too much, but when well made and served quickly, is one of the daintiest of desserts.

OMELET A LA DUCHESSE

This is a sweet baked omelet, and is served the same as one would serve an omelet souffle.

  6 eggs
1/2 cupful of water
1/2 a lemon's yellow rind, grated
1/2 cupful of thick cream
1/2 cupful of granulated sugar
  1 teaspoonful of vanilla or orange flower water
  1 small bit of cinnamon

Put the sugar, water, cinnamon and lemon rind over the fire, boil until it spins a thread and stand aside to cool. Separate the eggs; beat the yolks until creamy, and add the cream, then the strained syrup. Add the vanilla, and when cool fold in the well-beaten whites. Turn at once into a shallow silver or granite dish, dust thickly with powdered sugar and bake in a quick oven until brown.

OMELET SOUFFLE

This is, perhaps, one of the most difficult of all dishes to make. When, however, you have accomplished the art, you have one of the most satisfactory desserts. Like the preceding recipe, it must be made at the last moment and sent from the oven directly to the table. The eggs must be beaten to just the right point and the oven must be very hot. Get everything in readiness before beginning to make the souffle.

Select a bowl, perfectly clean, and arrange the star tube and pastry bag, if you are going to use one. If not, get out a baking dish. Sift six tablespoonfuls of powdered sugar. Separate six eggs. Put three of the yolks aside (as you will only use three), and beat the other three until creamy. Beat the whites until they are very stiff but not dry or broken. Now add three tablespoonfuls of the sifted powdered sugar. Beat for fully ten minutes. Then add the beaten yolks, the grated rind of a lemon and at the last a tablespoonful of lemon juice. Mix carefully and quickly, but thoroughly. Put four or five tablespoonfuls of this in the bottom of a platter, or baking dish. Put the remaining quantity quickly in the pastry bag, and press it out into roses. It is easier to make it in small rosettes all over the foundation. Dust quickly with the remaining three tablespoonfuls of sugar. Bake in a quick oven until golden brown. This will take about five minutes. Serve immediately. To be just right, this must be hot to the very center, crisp on top, moist underneath. If baked too long, the moment the top is touched it will fall, becoming stringy and unpalatable.

Omelet souffles are frequently flavored with rum, which must be mixed with the sugar. Sometimes they are sprayed with sherry just as they are taken from the oven. They may be built up into different forms, and garnished with candied or maraschino cherries, or chopped nuts.

www.ingramcontent.com/pod-product-compliance
Lightning Source LLC
Chambersburg PA
CBHW050020230526
45470CB00003B/1062